MW00974744

Not All Superheroes Wear Capes

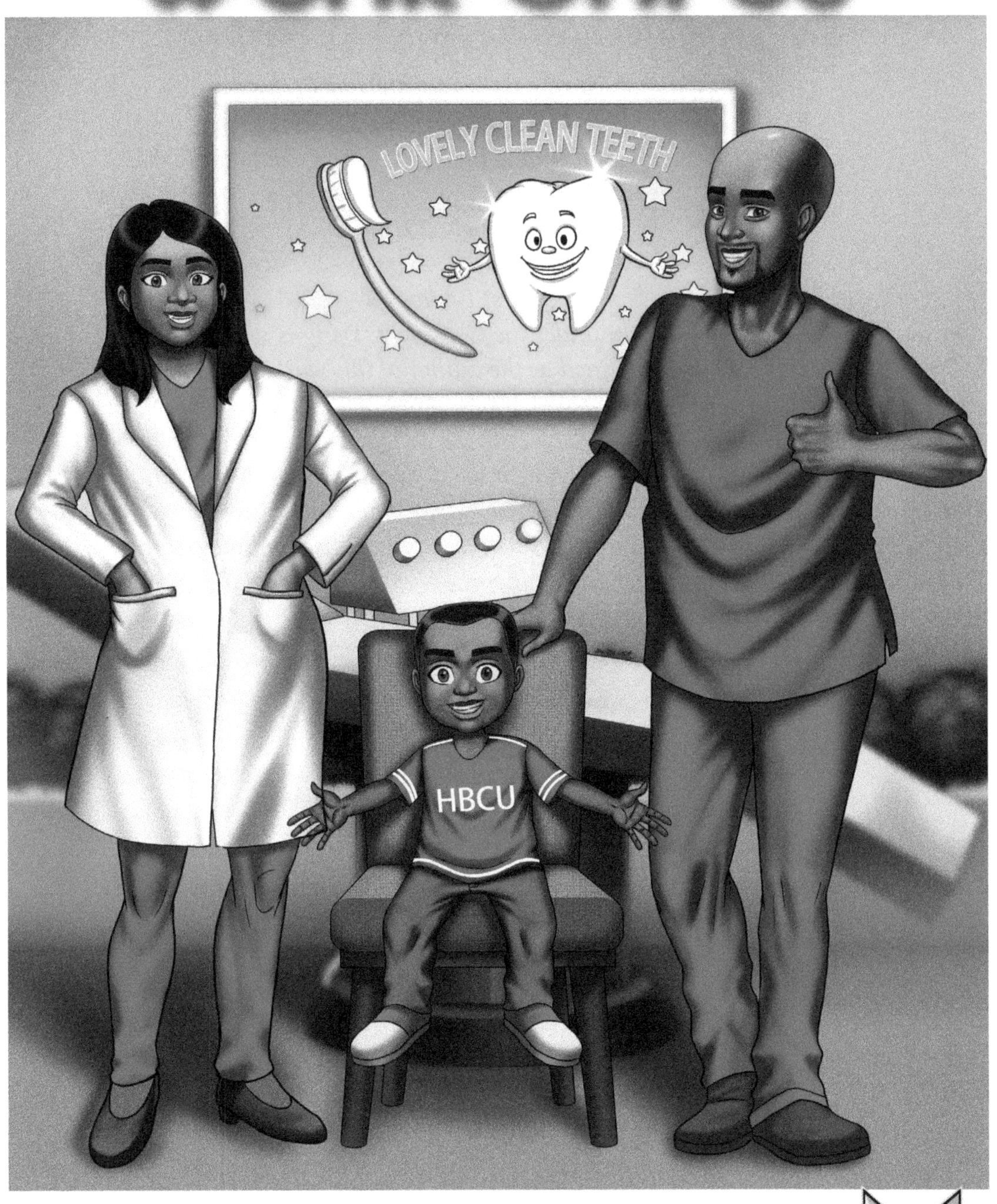

Written by: Alecia Heffner

Melanin Origins

Published by Melanin Origins LLC
PO Box 122123; Arlington, TX 76012

First Edition

Library of Congress Control Number: 2018939660

ISBN: 978-1-62676-769-0 hardback
ISBN: 978-1-62676-768-3 paperback
ISBN: 978-1-62676-767-6 ebook

Dedication

This book is dedicated to my son Ashton. Although life is just beginning for you, know that you are a king and have a unique place in this world. I hope that you dream big and break all barriers. Also, this book is particularly special as I wish to spark the interest of African-American children to see themselves in careers that they may not have otherwise imagined. Positive imagery is long overdue. It is my hope that this book inspires you to see people that look like you achieve their goals, despite being a part of a marginalized group. My niece Amani, I am excited that you have a desire to pursue medicine. I know you will persist. Representation matters! Countless educators have inspired me to truly value education, which is the basis of my book. Education will open doors. Thank you Mrs. Martin-Morgan, Dr. Singleton, Dr. Clark, and Dr. Epps.

Acknowledgment

My family has played a vital role in bringing my vision to life, thank you all. My late father's entrepreneurial spirit (Oscar), and my mother's (JoNell) commitment to being a healthcare provider for over 30 years motivated me to influence others in a positive way. We often become what we see. To my sisters, (Andrea and Alanna) thank you for your sacrifices, encouragement, saving graces and input. The spirit of my brother (Oscar) has been with me, as I have stayed up tirelessly working through the writing, illustration, and marketing process. To my nephews, be great in all that you do (Malcolm, Martell, Marquis, Chase, Davon, Jalen, and Chase David). I can't forget my eldest brother Cedric, I love you.

Lastly but not least, my heartfelt gratitude goes out to several special individuals and causes. Mark, thank you for your support throughout this entire process. You are a blessing most of all, but also a role model. Young African-American boys need more examples of positive, hardworking, educated African American men in a society where they are endangered. You embody all of this; you have exceled in STEM and continue to reach higher heights in your career. I appreciate you for riding with me on this journey. This is one of many successes that we will embark upon. Courtney, for always being there to push me to make things happen, and give unfiltered advice, our friendship is forever. Lawrence, for helping me to narrow down my title, and being there on my many pursuits. The Innocent Brown Girl Project. Mom's of NOLA. My Northeastern University doctoral family (NEU.POC.COP), Terrence, for keeping me grounded in my studies. Ashley, Lashawnte, Tashi, Amy, Carina, Marcus, Samantha, Stephanie, Sheremetria, no matter the distance, each of you are always looking out. My beloved alma mater, Fair Dillard. HBCU's forever!

Blessings,
Alecia

Superheroes come in many different shapes and sizes,
They stop the bad guys and prevent nasty surprises.
Whenever trouble decides to start heading your way,
You can rest assured that they will always save the day!

The superheroes of stories told often wear bright capes
And they drive cars or fly planes that have a special shape.
Yet, the best superheroes that our world always gets to see
Are the people who really look just like you and me.

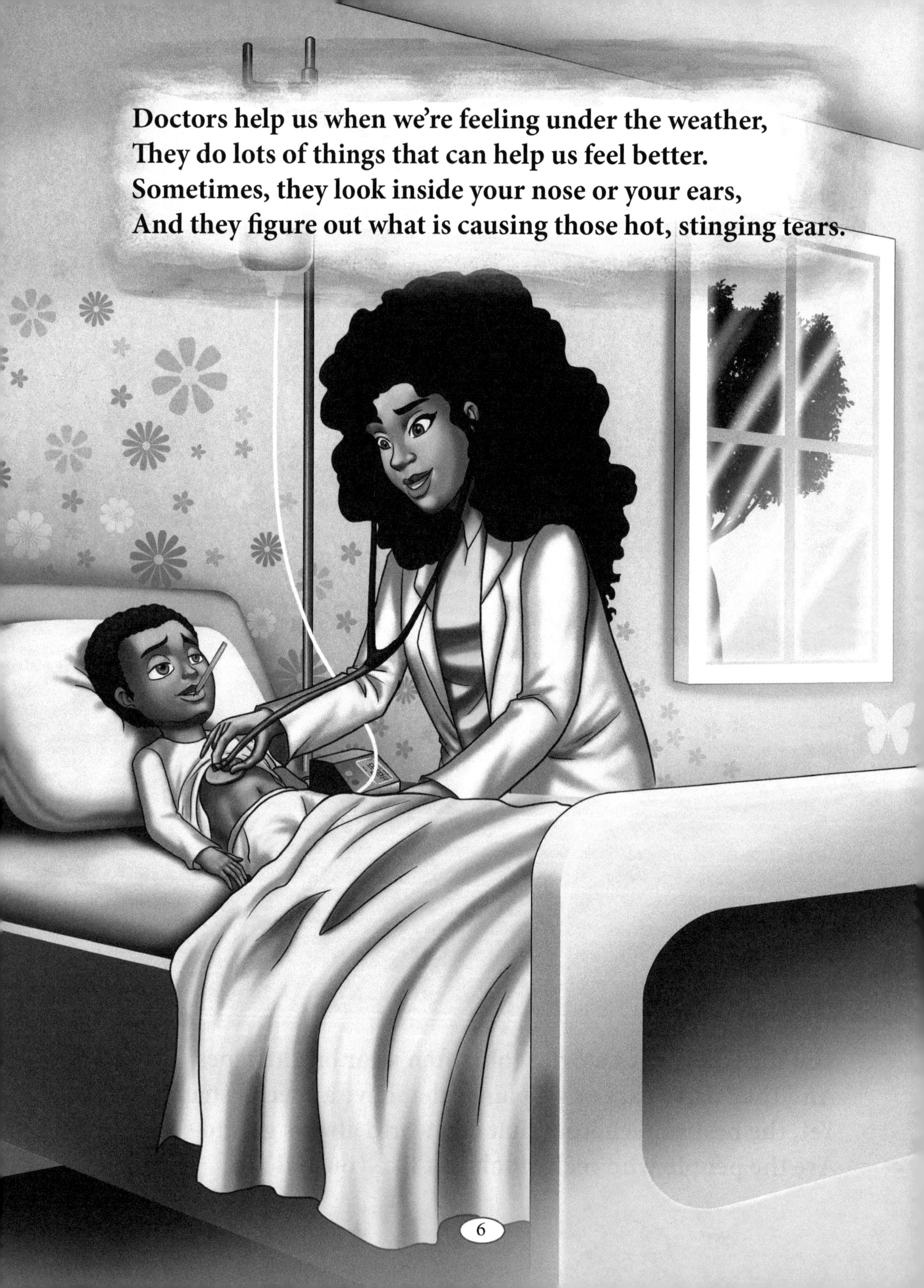
Doctors help us when we're feeling under the weather,
They do lots of things that can help us feel better.
Sometimes, they look inside your nose or your ears,
And they figure out what is causing those hot, stinging tears.

Some of them work with kids and that is all they do,
Others help people with specific illnesses make it through.
From fevers to tummy aches and everything in-between,
Doctors are some of the best superheroes ever seen!

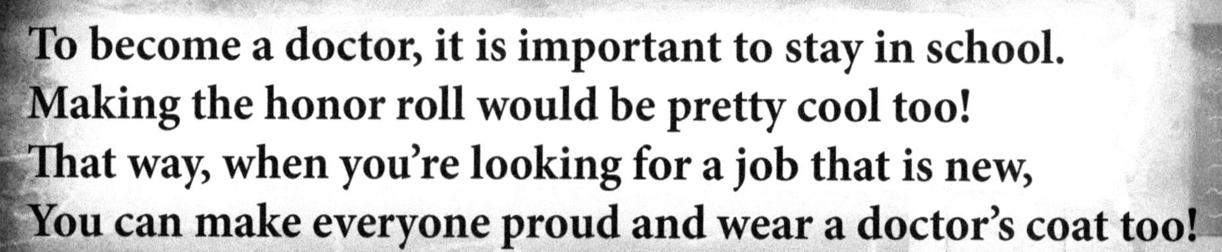

To become a doctor, it is important to stay in school.
Making the honor roll would be pretty cool too!
That way, when you're looking for a job that is new,
You can make everyone proud and wear a doctor's coat too!

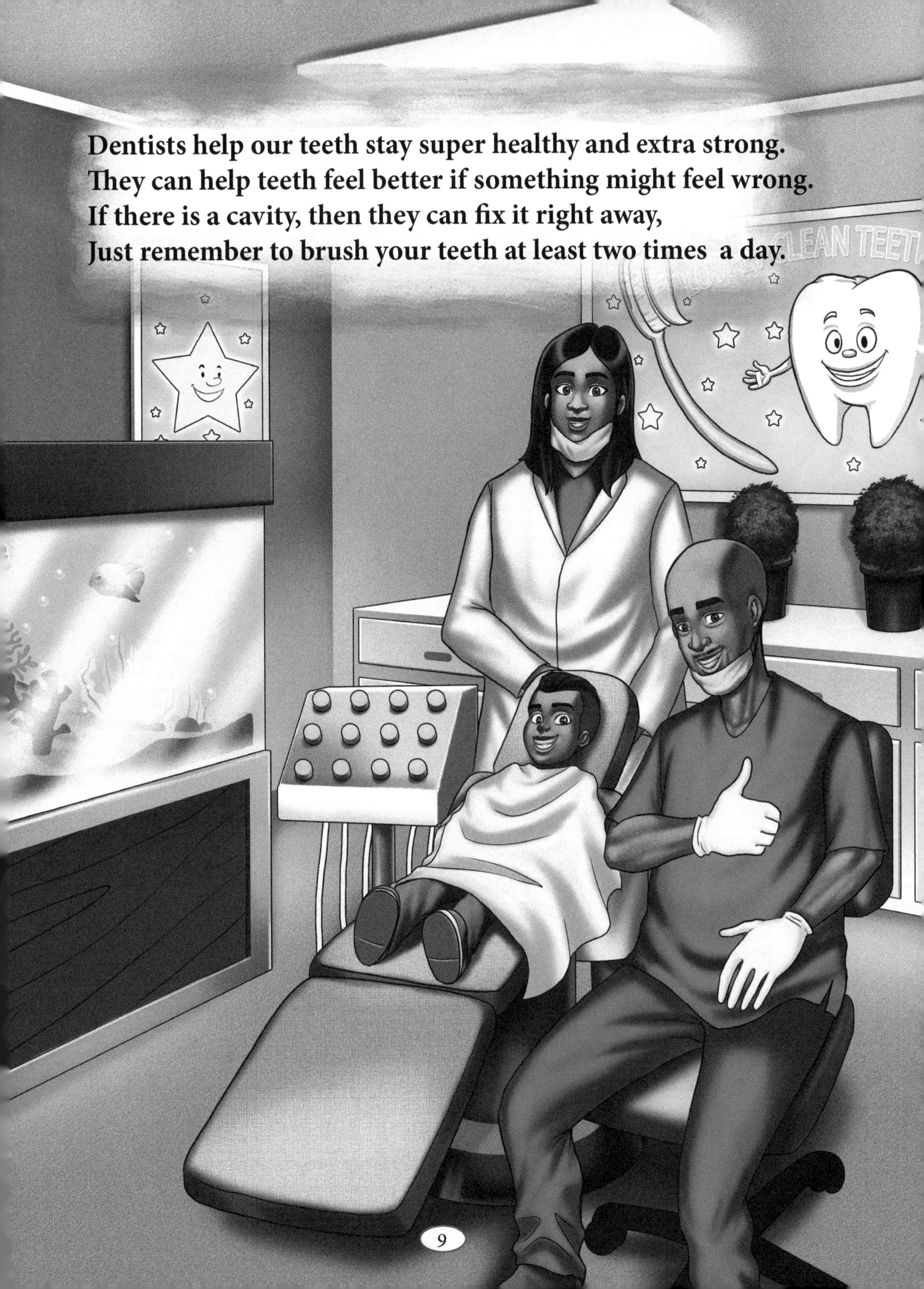

Dentists help our teeth stay super healthy and extra strong.
They can help teeth feel better if something might feel wrong.
If there is a cavity, then they can fix it right away,
Just remember to brush your teeth at least two times a day.

Then the gums are checked to ensure a healthy pink color.
And special sealants help to stop teeth from growing duller.
It's also a chance to get a smile that's super super clean.
Dentists are some of the best superheroes ever seen!

To become a dentist, it's important to study hard.
Making good grades will become your safeguard.
That way, when you're looking for a job that's new,
You can make everyone proud and wear a dentist's coat too!

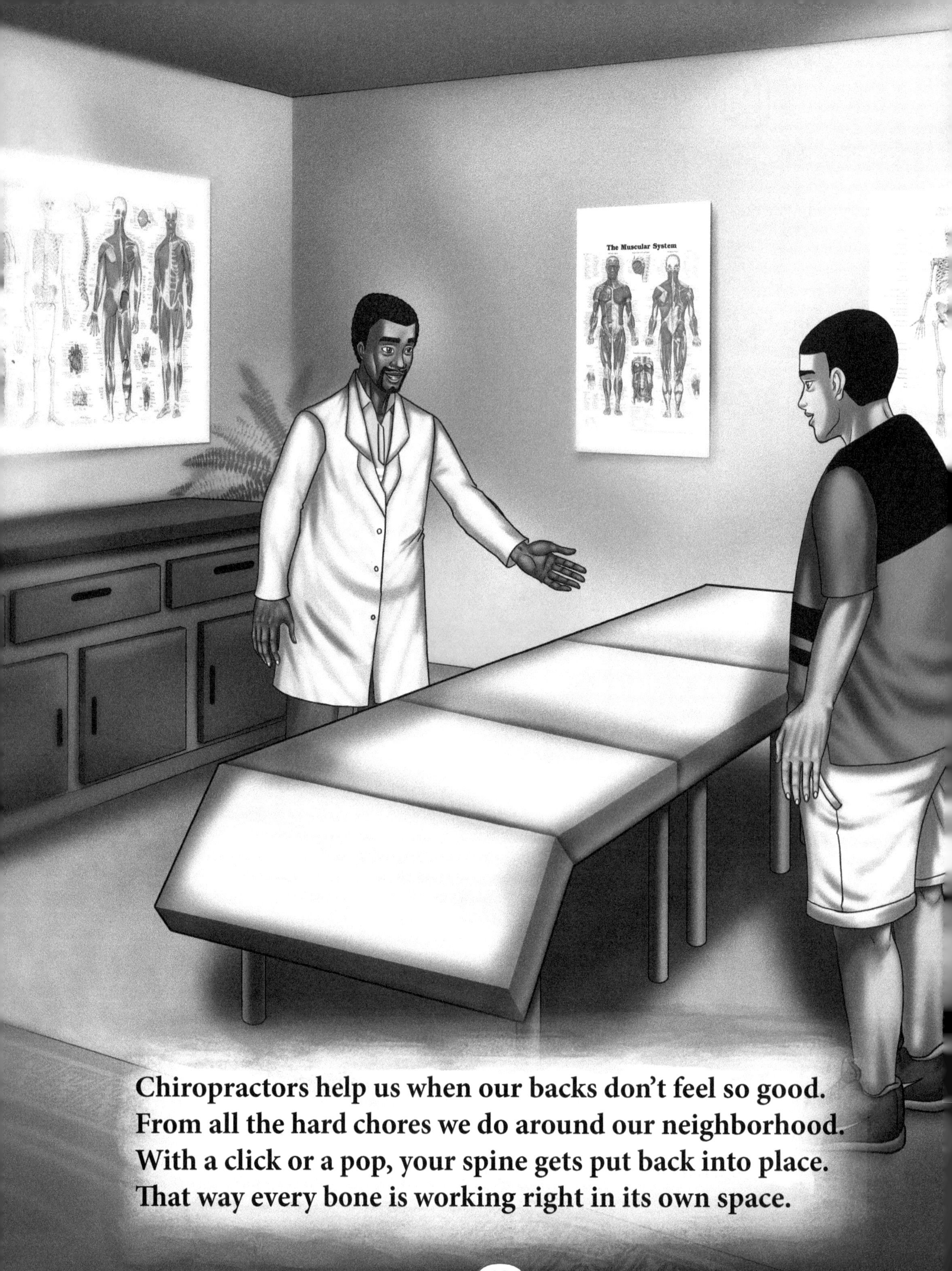

**Chiropractors help us when our backs don't feel so good.
From all the hard chores we do around our neighborhood.
With a click or a pop, your spine gets put back into place.
That way every bone is working right in its own space.**

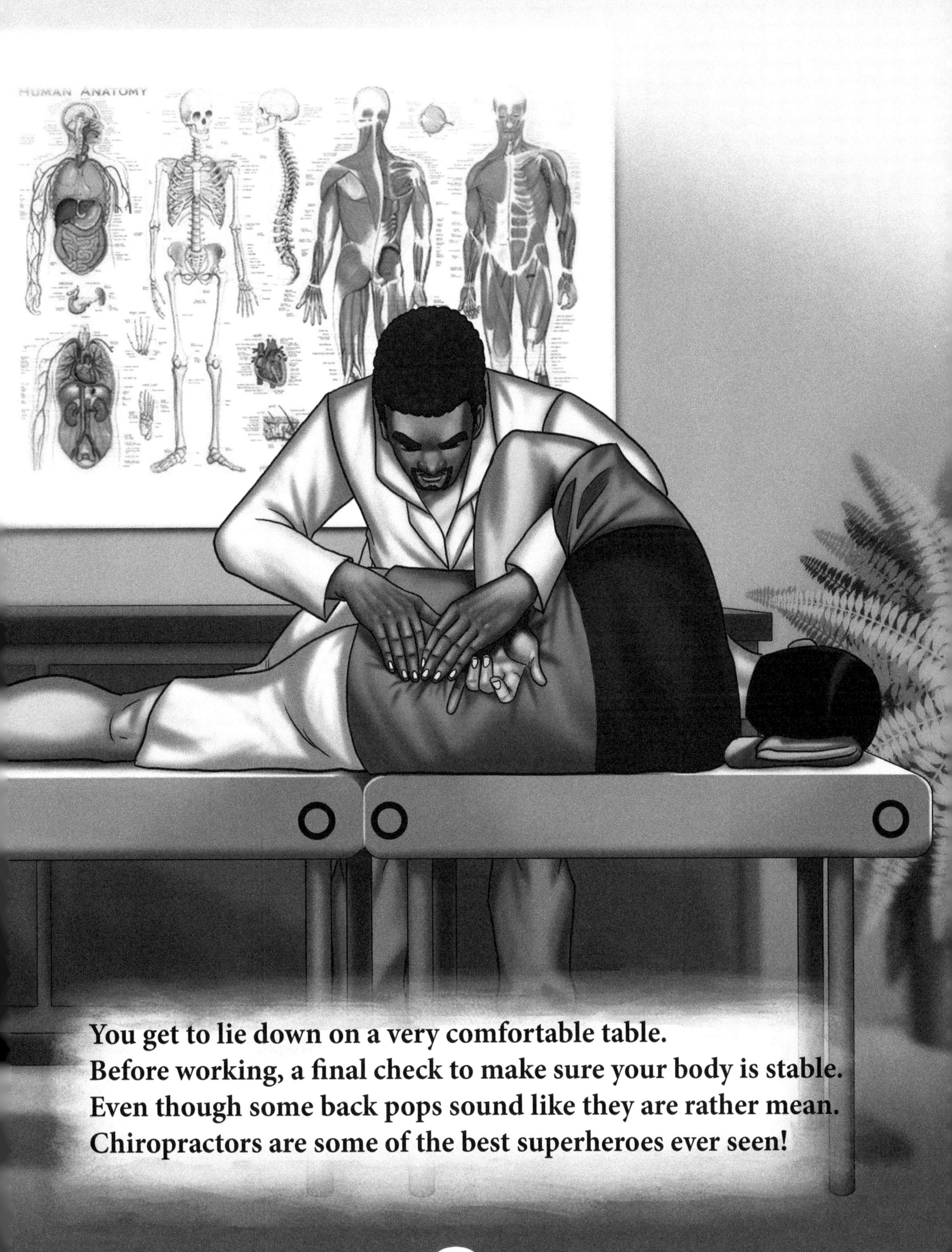

You get to lie down on a very comfortable table.
Before working, a final check to make sure your body is stable.
Even though some back pops sound like they are rather mean.
Chiropractors are some of the best superheroes ever seen!

To become a Chiropractor, it is important to do well in science.
Math is another subject upon which you'll have great reliance.
That way when you're looking for a job that is new.
You can make everyone proud and wear a Chiropractor's coat too!

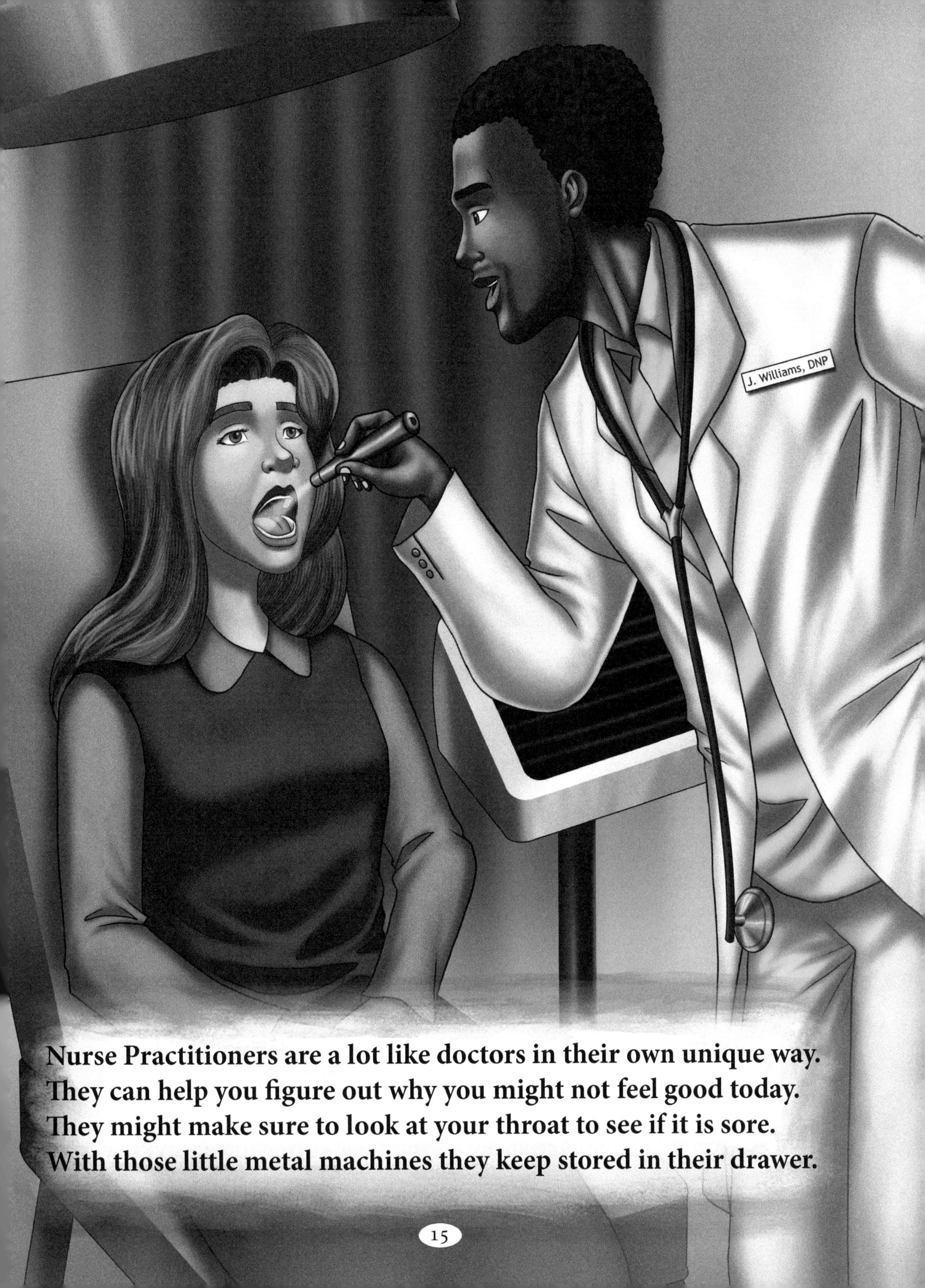

Nurse Practitioners are a lot like doctors in their own unique way.
They can help you figure out why you might not feel good today.
They might make sure to look at your throat to see if it is sore.
With those little metal machines they keep stored in their drawer.

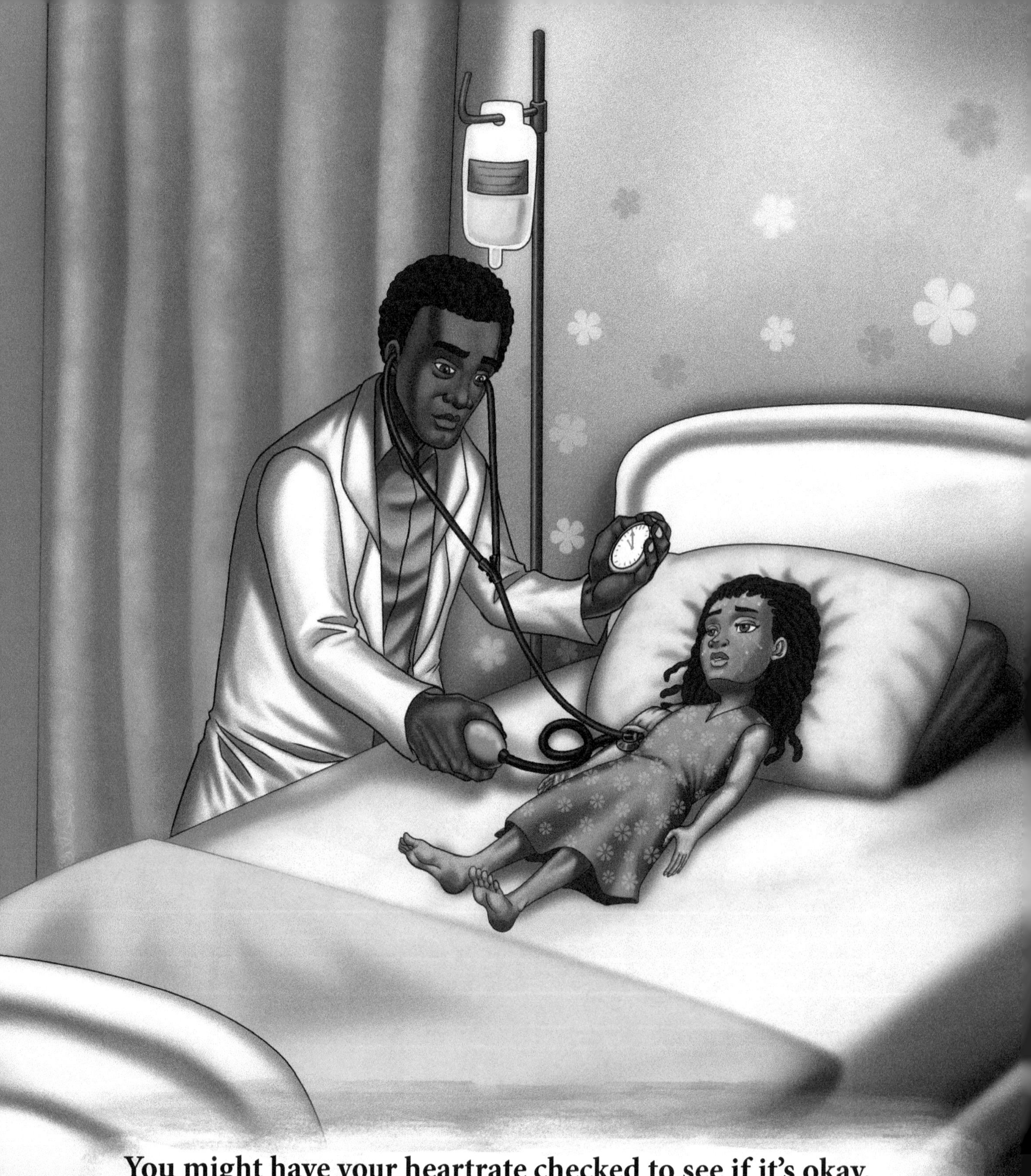

You might have your heartrate checked to see if it's okay.
And that big pressure cuff might squeeze you tight in its way.
When you don't feel good and you're looking a little green,
Nurse Practitioners are some of the best superheroes ever seen!

To become a Nurse Practitioner, you'll need to study long.
You'll need to graduate with grades that are super strong.
That way, when you're looking for a job that is new.
You can make everyone proud and wear a Nurse Practitioner's coat too!

Physical Therapists help us feel better by showing us ways to exercise. Especially after a big OUCHIE happens that brings tears to our eyes. Each exercise works where it hurts, helping the spot become strong. Some exercises are short, but others can be quite long.

You might even be asked to do some exercises at home too.
That way, you can once again become the very best you.
Everyone can visit, whether you're young, old, or in between,
Physical Therapists are some of the best superheroes ever seen!

To become a Physical Therapist, you'll need to have what it takes.
There are no shortcuts here and no room for fakes.
That way, when you're looking for a job that's new,
You can make everyone proud and wear a Physical Therapist's coat too!

PHARMACY

Pharmacists give us medicine to help us feel better.
On a cold day, you might find them behind the counter wearing a sweater.
They take your prescription, then make sure you get the medicine that is right.
Then they talk to you to make sure you know how to take it, day and night.

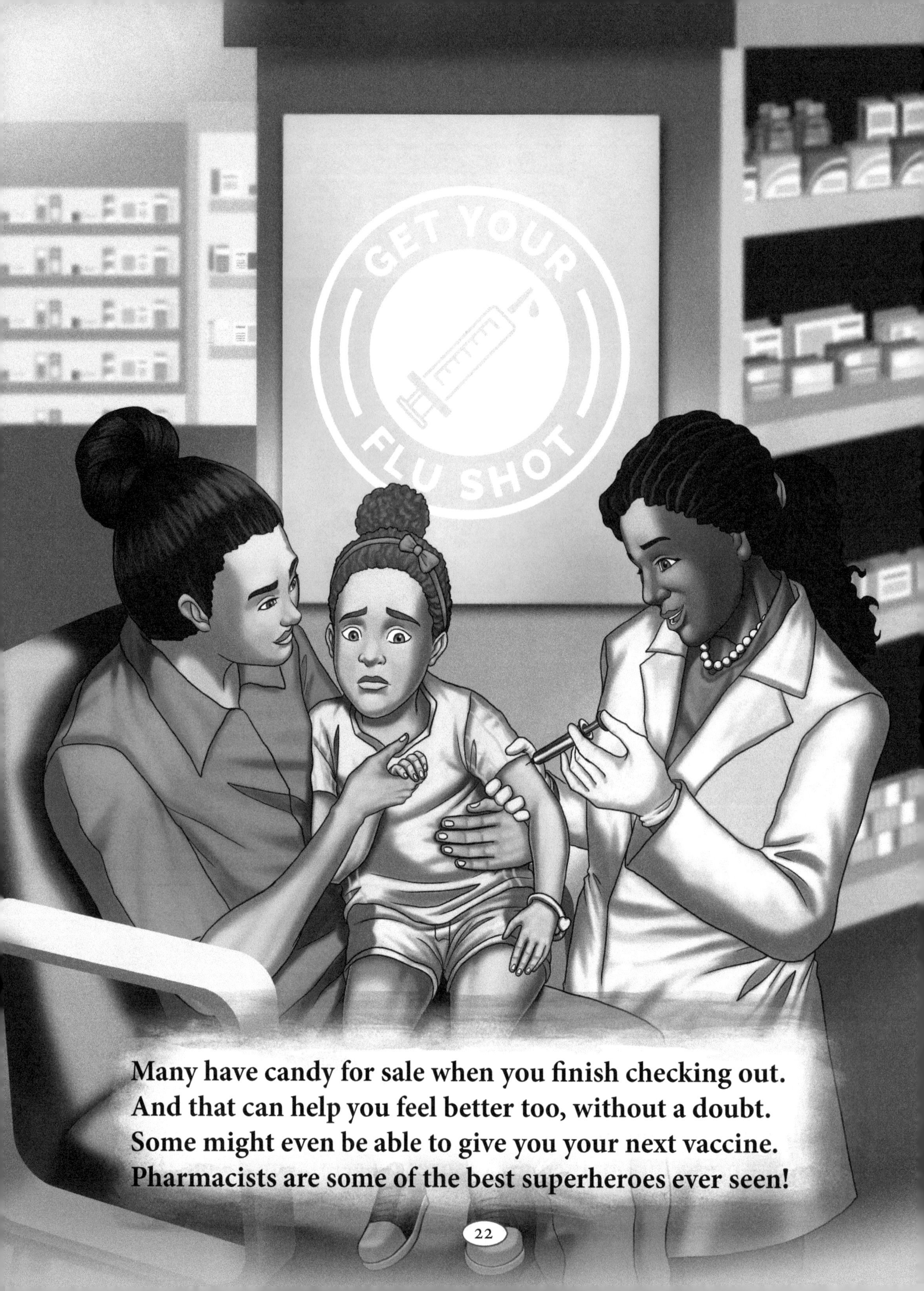

**Many have candy for sale when you finish checking out.
And that can help you feel better too, without a doubt.
Some might even be able to give you your next vaccine.
Pharmacists are some of the best superheroes ever seen!**

To become a Pharmacist, you'll need to learn about medicine, for sure.
You'll need to know the best ones, which might offer a cure.
That way, when you're looking for a job that is new,
You can make everyone proud and wear a Pharmacist's coat too!

Podiatrists are doctors that help our strong and healthy feet.
They look at our toes and ankles, so we can walk down the street.
If it hurts to walk or there is pain lurking at the bottom,
They'll work to figure out where exactly is the foot problem.

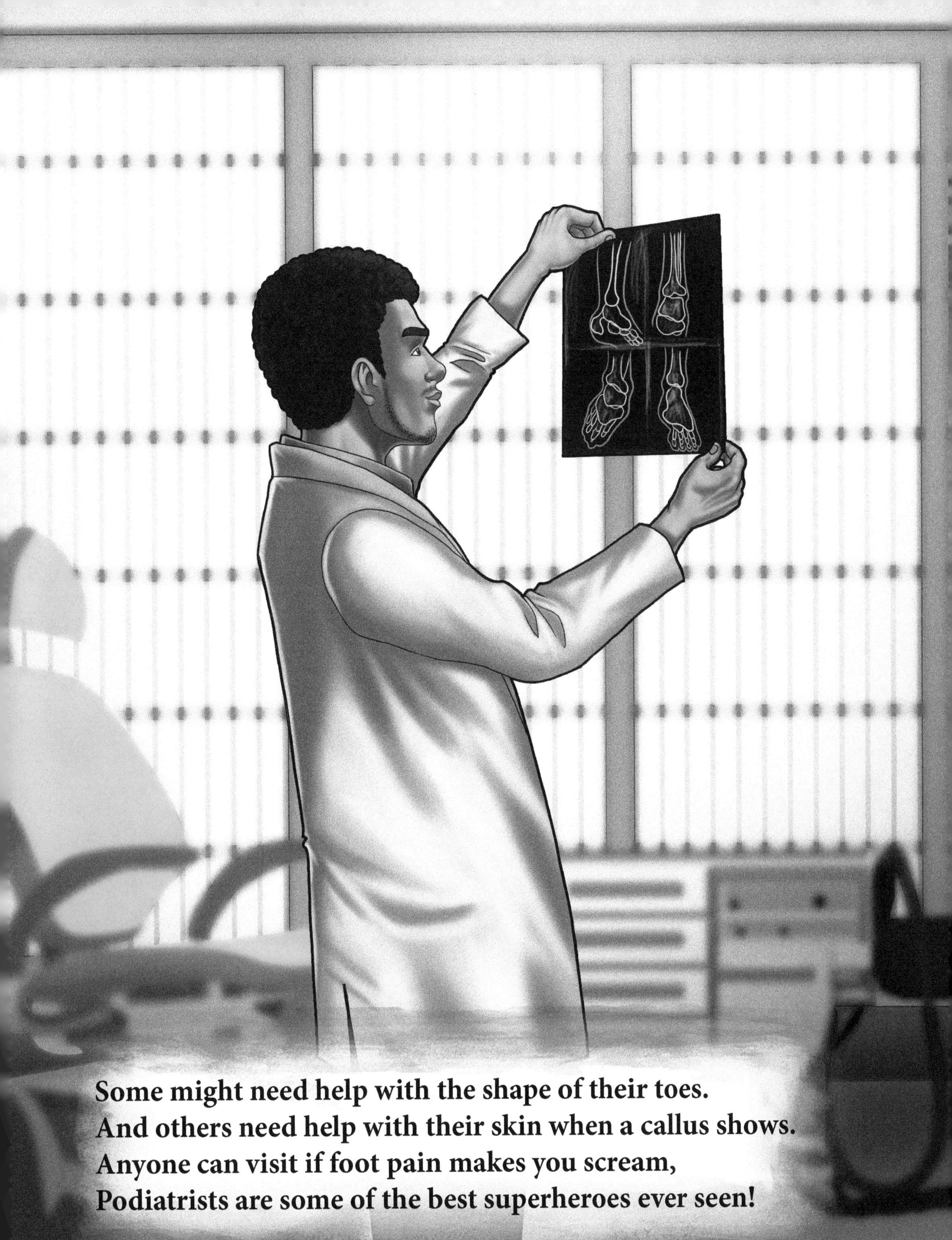

Some might need help with the shape of their toes.
And others need help with their skin when a callus shows.
Anyone can visit if foot pain makes you scream,
Podiatrists are some of the best superheroes ever seen!

To become a Podiatrist, you'll need to love looking at feet.
And doing well in school is something that can't be beat.
That way, when you're looking for a job that's new.
You can make everyone proud and wear a Podiatrist's coat too!

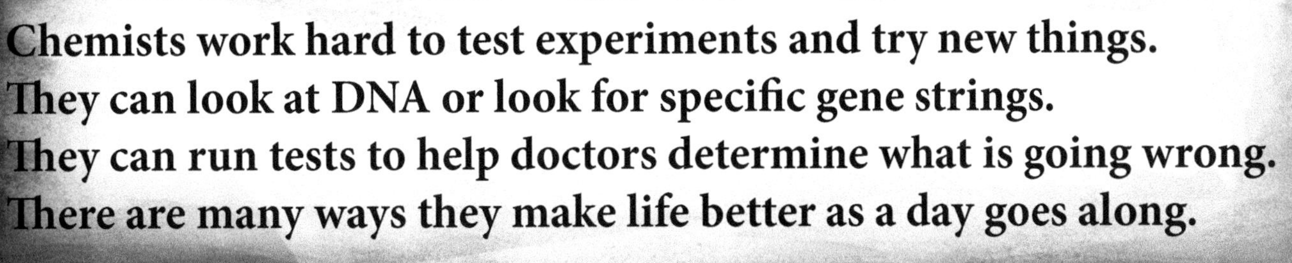

Chemists work hard to test experiments and try new things.
They can look at DNA or look for specific gene strings.
They can run tests to help doctors determine what is going wrong.
There are many ways they make life better as a day goes along.

To become a Chemist, there are many sciences from which to choose.
Following what you love creates a dream where you cannot lose.
That way, when you're looking for a job that's new,
You can make everyone proud and wear a Chemist's coat too!

Superheroes are all around us every day.
They help us feel better when good feelings go away.
They wear big white coats instead of a colorful cape.
And that's what helps to keep us in our best-ever shape.

It can be tough to become a superhero today.
But everyone has the chance to find their own way.
So strap on your shoes and work as hard as you can.
And good grades in school is for sure the greatest plan!

The one who stays prepared, and focused on their goals,
And never gives up on themselves, no matter what challenge beholds,
Becomes the best that they can be, and gains superhero knowledge,
And knows to be a real superhero, they must go to college!

CPSIA information can be obtained
at www.ICGtesting.com
Printed in the USA
BVHW011415101218
535227BV00004B/70/P

9 781626 767690